Building Your Own Arduino Shields: Interfacing with the Arduino Using Basic Components

by

David Leithauser

Table of Contents

Introduction

Arduinos are fascinating and versatile electronic components. They can be used to control a wide range of devices in response to a variety of inputs, provided that the proper electronic add-ons are connected to the inputs and outputs. There are many off-the-shelf gadgets that you can buy for Arduinos to connect to the inputs and outputs. However, there are two problems with relying totally on these off-the-shelf gadgets.

First, devices built by third parties can be expensive. If you build your own devices to interface with the Arduino out of the basic electronic components (resistors, transistors, etc.), you can save a considerable amount of money. You can easily save enough on one project to pay for this book. Cutting the cost of your projects is even more important if you hope to market your Arduino based gadget, since every penny you save on construction costs can be passed on to your customers and lower the retail price of your product.

Second, when you buy commercial add-ons to the Arduino, you are limited to what someone else has decided is worth doing with the Arduino. You might have trouble finding a device that handles as much current as you want, or has a fast enough response time, or otherwise fits the needs of your project. If you built your own interfaces, you can have the Arduino do exactly what you want it to in exactly the way you want it to.

This book explains in considerable detail how to use elementary electronic components to connect to your Arduino. It is written for someone who is reasonably comfortable with working with electronics (a reasonable assumption for someone who is building things with Arduinos). For example, I do not get into soldering techniques. I do, however, assume that the reader does not know enough about fundamental electronic components like transistors to design circuits from scratch. I will start off in Chapter 1 explaining a little about Arduinos and also showing the electronic symbols that I will be using in the diagrams in the book. In the following chapters, I will explain how to accomplish various general processes, such as measuring electrical resistance and voltage and controlling high output voltage and current from Arduino output pins. These may seem rather abstract at first glance, but they are actually what you need to understand in order to build your own input and output devices for the Arduino, as will be explained in the following chapters.

There are several differences between this book and most of the books you can buy about Arduinos. Most books on Arduinos these days are collections of fun (at least in the mind of the authors) and/or useful projects for you to do. The problem with this is that there is not enough information in most of these books to generalize to other projects, or even to modify the project being described. The author gives you lists of specific (usually off-the-shelf) devices to connect to the Arduino to accomplish a specific project, with very little background. You learn how to do a few things, but do not come away with a real understanding of how to do things in general. This book does not have complete projects, although a few of the designs can be used by themselves as voltmeters or other electronic test instruments. The purpose of this book is to teach general design, not specific projects. The second difference is that this book contains some mathematical equations. Don't worry, I will not be getting into a lot of heavy or theoretical math. The equations in this book are just some formulas for selecting the right size components such as resistors. This is the secret to allowing you to generalize the schematics in this book. I will give you the plans for general circuits for doing widely useful functions that will allow you to then create specific projects you want, and give you the means to select the components you need to do that.

Chapter 1

A quick primer in Arduinos and electronics

First, you need to know a few details about the Arduino, especially its limitations. The Arduino is actually a specialized computer. It is a control device. Instead of accepting input from a keyboard or mouse and provide information to a screen, it is designed to read various analog and digital electrical inputs and provide various electrical outputs. The actual Arduino circuit usually runs on 5 volts. Some can run on 3.3 volts, but we will assume in this book that the Arduino is running on 5 volts unless otherwise stated. However, the Arduino board has a voltage regulator that reduces the voltage from your power supply down to 5 or 3.3 volts, so you can power the Arduino board with up to 12 volts (most models) or 23 (for the BUONO). Just be sure to apply anything over 5 volts to the power socket, not to one of the input pin holes. Applying the power supply directly to one of the pins would bypass the voltage regulator and damage your Arduino.

For analog inputs (usually A0 to A5), the Arduino can accept a voltage from 0 to 5 volts relative to the ground (GND) pin. Internally, the analog pin generates a number that is 0 for 0 volts input and 1023 for 5 volts. That is, the reading at an analog pin is approximately 204.6 times the voltage applied. I say approximately because the number is always an integer, so fractions are truncated (rounded down). Thus, for example, 1 volt would be reported as 204, 2 volts as 409, and 2.5 volts would be reported as 511. Note that you must be careful not to apply a negative voltage or a voltage more than 5 to the analog input pin.

Digital inputs return either LOW or HIGH (These are reserved keywords). A digital input pin returns LOW if the voltage applied to it is 2 volts or less. It returns a value of HIGH if the voltage applied to it is 3 volts or more. Between 2 and 3 volts input, the reading is unstable. One subject we will discuss in this book is buffering the input for digital pins so it is always either below 2 or above 3 volts.

Digital outputs output 0 (the same as the ground connection) if they have been told by programming to be LOW. They output 5 volts if they are programmed to be HIGH. There is no actual analog output that gives a variable analog value from 0 to 5 volts, but there is a rough simulation. Some digital outputs can be configured to output a signal that jumps from 0 to 5 volts and back to 0 very rapidly (about 500 times per second). This is called Pulse Width Modulation, and pins that can do this are labeled PWM on the Arduino. You can control what percentage of the time this signal is at 0 volts and at 5 volts. You do this by using the analogWrite(outputpin,N) command in your code, where outputpin is the number of the digital output pin and N is a number from 0 to 255. If N is 0, the output is 0 all the time. If N is 255, it is 5 volts all the time. Between 0 and 255, output is 5 volts a fraction of the time determined by N/255. For example, if N is 100, the output is 5 volts 100/255 of the time. We will discuss in this book converting this to a true analog output.

In this book, we will discuss using very common and inexpensive components to interface with the Arduino. Typical components are resistors, capacitors, transistors, op-amps, and other parts that you can buy for a few cents (or at most, a few dollars) apiece on EBay, Amazon.com, or other discount stores. This book contains circuit diagrams using standard symbols. These are shown in Figure 1.1.

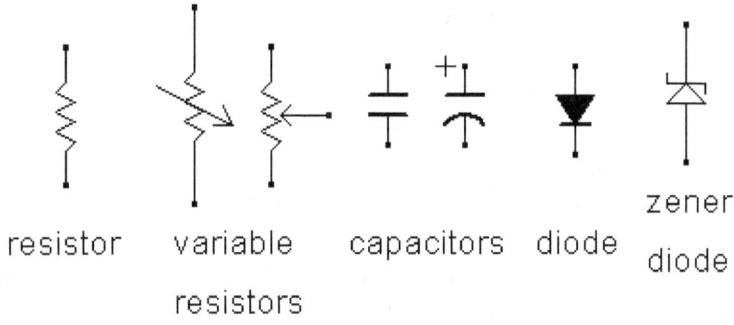

resistor   variable   capacitors diode   zener diode

resistors

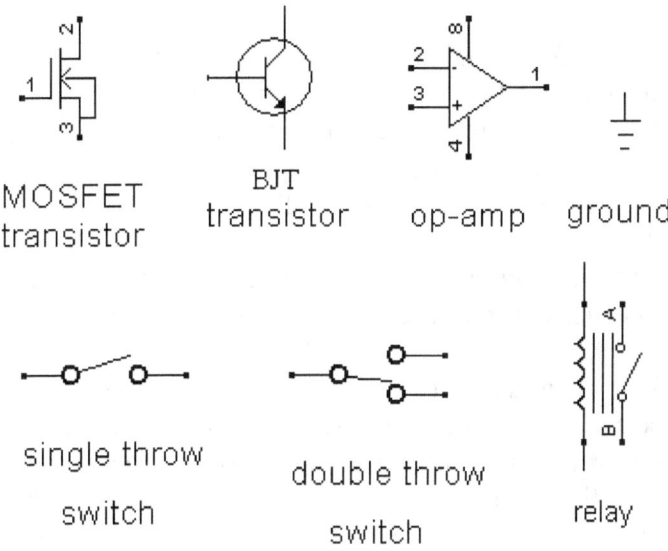

MOSFET transistor   BJT transistor   op-amp   ground

single throw switch   double throw switch   relay

Figure 1.1

You may notice that some of the components have several symbols. There are two symbols for variable resistors. The one on the right is usually used when you are connecting to three points on the variable resistor, while the one on the left is usually used when you are only using one end connection and the center tap. I will sometimes use the one on the right in this book to represent an unspecified resistance that varies due to the external environment, such as a photoresistor or a thermistor. For the capacitor, the symbol on the left is the more general symbol to represent

7

any kind of capacitor. The symbol on the right represents specifically an electrolytic capacitor, which can usually attain a higher capacitance value than other capacitors but must always have a positive charge on one side. The two symbols for transistors represent different types of transistors. They perform basically the same function (a voltage applied at one connection controls the current flow between two other connections), but the details are different. There are quite a few symbols that people use for ground, but I will always use the symbol for an earth ground for consistency, even if it is not technically the correct one for a circuit.

A normal diode is a component that only allows current to flow one way. This has the basic effect of converting alternating current (AC) into direct current (DC). A zener diode is a special diode that allows current to flow in one direction normally, but will allow current to flow in the opposite direction if the reverse voltage exceeds a certain amount. Thus, it can be used as a voltage limiter. That is, it limits the voltage that can be across the zener diode.

Op-amps (operational amplifiers) are very useful components. They have two inputs, the inverting and the non-inverting. The difference between the voltage at the inverting and non-inverting inputs is multiplied by a very large number and goes to the output. In fact, in the ideal op-amp, the multiplication factor is infinite. In reality, it is normally many thousands. In the most common use, some of the output is fed back into the inverting input creating a negative feedback loop to counteract the gain, thus bringing the amplification down to a reasonable number. Most op-amps are designed to have a dual power supply so that there are three power leads: V+, V-, and ground. For our purposes, we will want the op-amp to be powered by one source. Often this is the 5 volt pin of the Arduino, so that the power supply to the op-amp is just V+ (at 5 volts) and ground (the Arduino ground), but it can be a higher voltage (as explained in later chapters). There are some op-amps designed to work with a single power supply, several specifically to work with 5 V digital circuits. Two common (and inexpensive) ones are the LM358 and the NJM2904. On the op-amp symbol I use, the connection labeled 8 is the positive power supply connection and connection 4 is the negative (ground) power connection. Lead 2 is the inverting input and lead 3 is the non-inverting input. These numbers are used because these are the most common pins is the dual op-amp (two op-amps on one IC) ICs.

Regarding the transistor diagrams shown above, on the MOSFET, the connection labeled 1 is the gate, number 3 is the source, and 2 is the drain. This is for N channel MOSFETs, which is what we will be using in this book. On the BJT, the side contact in the diagram is called the base, the top contact is called the collector, and the bottom connection is called the emitter. This is for NPN type BJT transistors, which we will be using in this book. It is important to know this terminology, because you will need to be able to know what leads on the physical transistors correspond to the connections in the diagrams.

A few notes regarding resistors: Resistors are usually rated in ohms, kiloohms (1,000 ohms, abbreviated K), or megaohms (1,000,000 ohms, abbreviated M). Thus, for example, a 10,000 ohms resistor will be referred to in this book as 10 K. The accuracy of many of the analog measurements you get from the circuits in this book will depend on the accuracy of the resistors. Resistors are rated for how accurately the actual resistance of the resistor matches the advertised resistance. In poor quality resistors, the actual resistance can be as much as 20% different than the claimed resistance. For example a 1 K (1,000 ohms) resistor could actually be 1,200 ohms or 800 ohms. You should use resistors rated at no more than +/- 5% for the analog circuits in this book (mostly chapters 3 – 5), and preferably +/- 1% for these circuits. You can buy resistors rated at being accurate to within .1% or even .02%, but these are expensive. A cheaper alternative in the long run could be to use a high quality resistance meter to check the resistance of a resistor before using it in a circuit. For circuits where the resistors are merely protective current limiting resistors, the exact value is not important, and you can use cheaper resistors. Another feature of resistors is their power rating rating. The power flowing through a resistor is the current flowing through it times the voltage across it. In most (but not all) of the circuits in this book, the voltage across a resistor will be no more than 5 volts, and the current will be no more than 40 milliamps (thousands of an amp), so the power will be less than .25 watts, a common rating for resistors. However, to be on the safe side, I recommend using at least .5 watt resistors if possible.

Figure 1.2 shows the symbol I will use for the Arduino in my circuits. Not all Arduinos are the same or have the same number of connections (for example, some have more digital or analog inputs and outputs), so I will use a fairly low common denominator model, the Arduino Uno.

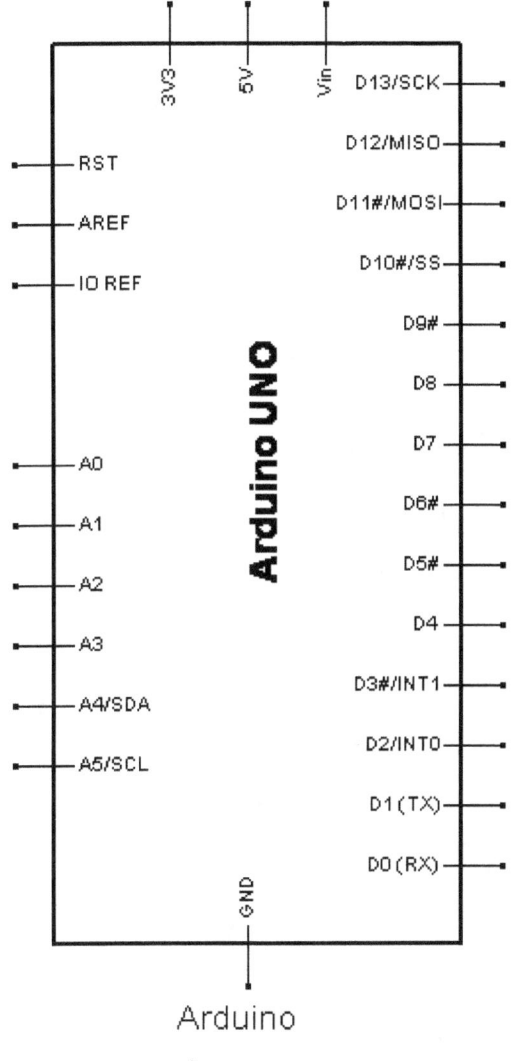

Arduino

Figure 1.2

Wires are shown as lines. In drawing diagrams, it is sometimes necessary to have wires cross but not be connected. To distinguish when wires are actually connected and not just crossing in the diagram, I will put a dot at the intersection when the wires are actually connected, as shown in Figure 1.3.

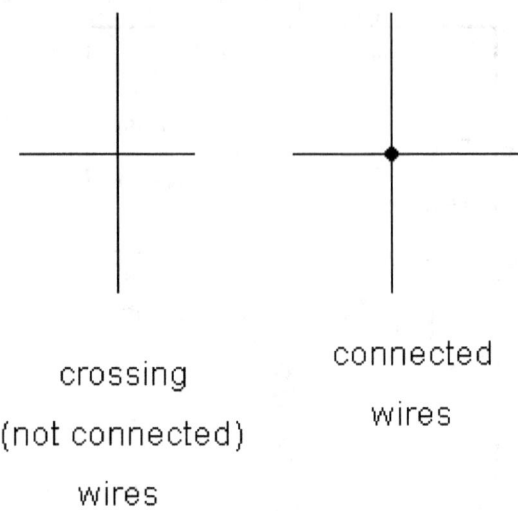

crossing
(not connected)
wires

connected
wires

Figure 1.3

Chapter 2

Protecting the Arduino from dangerous input and output conditions.

Although Arduinos are inexpensive, you still do not want to damage them with unusual electric conditions, such as accidental shorts that draw too much current or applying too much voltage. Besides the expense of replacing the Arduino, there is the trouble of rebuilding the device and any inconvenience you suffer due to having the device off line until it is repaired. So, before going into specific circuits, I will provide a few general tips in this chapter to protecting your Arduino from overloads and other problems.

One useful component is the zener diode. At low voltages, this component allows current to flow in one direction and blocks it in the other, like any diode. However, a zener diode will allow current to flow freely in the reverse direct (the direction it normally blocks current) if the voltage exceeds a certain level. The level is controlled during manufacture and is set. Zener diodes come in various voltage levels. Since Arduinos almost always operate at either 5 volts or 3.3 volts, these are the levels we are interested in. Fortunately, zener diodes are commonly available in 5.1 V and 3.3 V. If you put a zener diode in series with a resistor, the voltage across the zener diode will not exceed the voltage rating of the zener diode. Figure 2.1 shows a general circuit that protects against excess voltage at an Arduino pin.

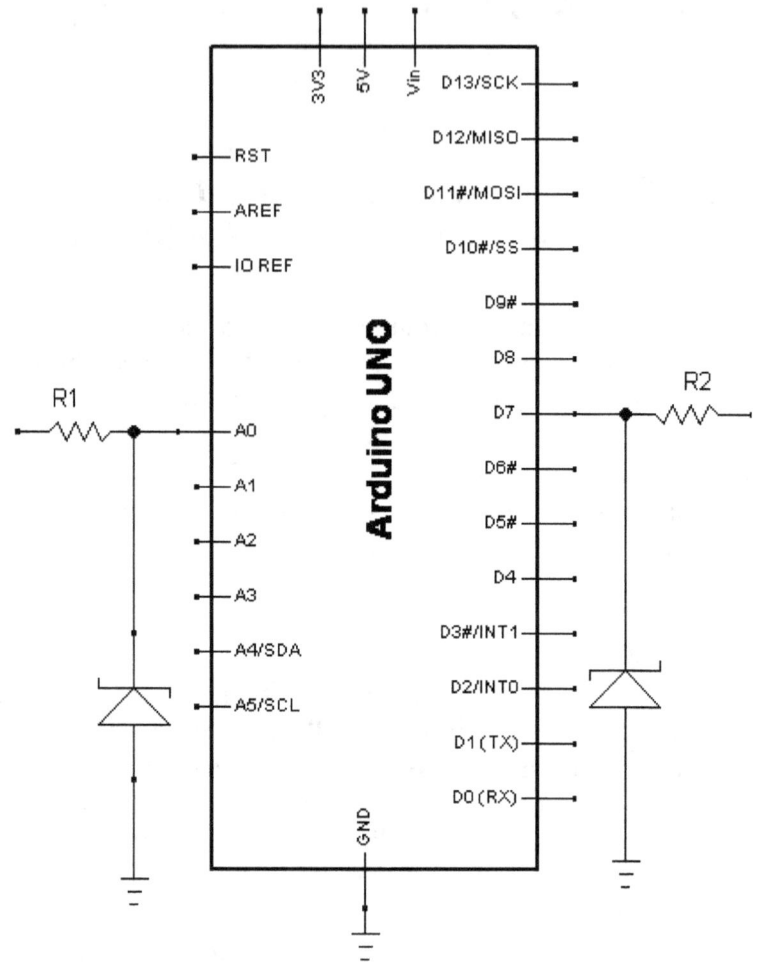

Figure 2.1

If the input voltage goes above the limit for the zener diode, it starts conducing current and a counter voltage appears across the resistors, limiting the voltage to the limit voltage of the zener diodes. The actual values of the resistors are not too important, but they should be fairly high to limit the current that flows through the zener diode. Because the Arduino has a very high input resistance, the resistors will have very little effect on the input. I would suggest values of about 10 K for the resistors. Note that you would normally only need to do this if the Arduino pins (A0 and D7, in this case) are configured for input, and it is possible for the inputs to exceed the rated voltage of the Arduino input pins. However, there are some cases where you might need to protect output pins, as I will discuss shortly.

For output pins, there are several concerns. One is the possibility of drawing too much current. This can happen if the component you are sending the output signal to shorts out, or simply has a lower resistance than you expected. Some components have a cascading effect, where under some circumstances they can suddenly start drawing a great deal of current. The simplest solution to that is to provide a current limiting resistor to the output pin, as shown in Figure 2.2.

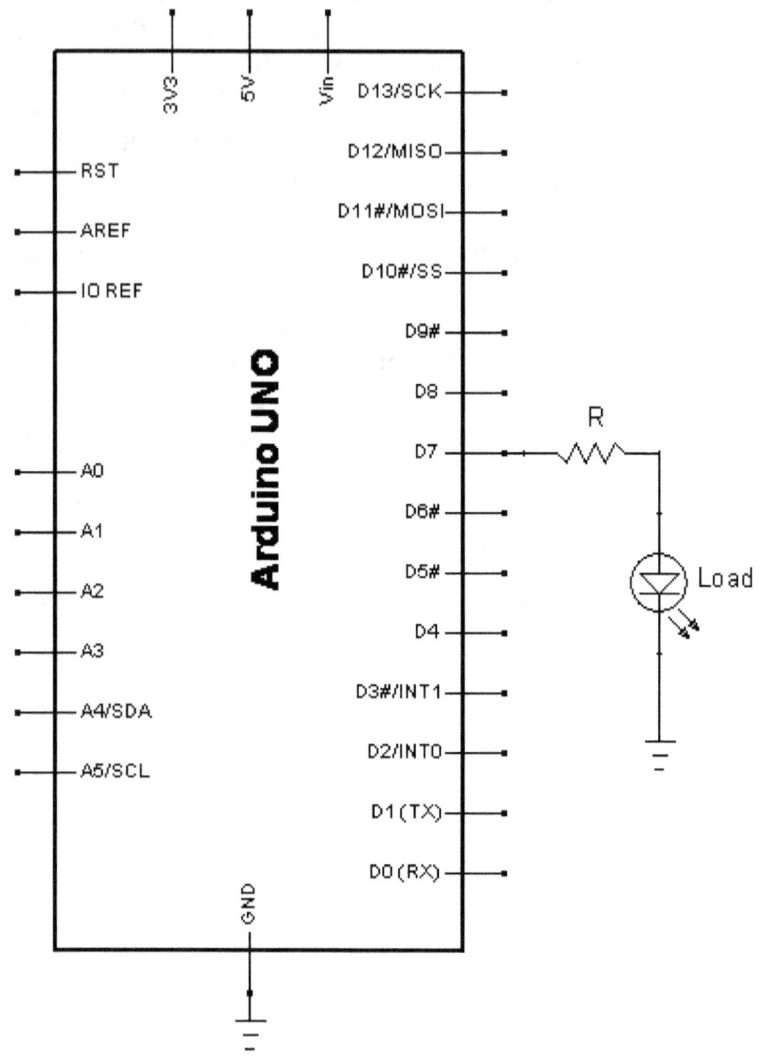

Figure 2.2

For the illustration, I used an LED to demonstrate the load, but the load can be anything. The disadvantage of using a resistor in this situation is that if the load is drawing a fairly high current, there will be a voltage drop across the resistor, thus slightly reducing the voltage the load. You therefore want as low a value as you can have for the resistor and still limit the current to a safe level in case of a short or similar problem. The maximum safe output current for most Arduinos is 40 mA. At 5 V output, this would mean you need a minimum of 125 ohms. I would recommend a minimum of 150 or 220 ohms. For most practical purposes, you can use 1 K.

Some loads that have inductance, such as relays, motors, or speakers, can produce a voltage spike when the voltage across them is suddenly cut off or changed. To protect against this, you should put a diode across the coil as shown in Figure 2.3.

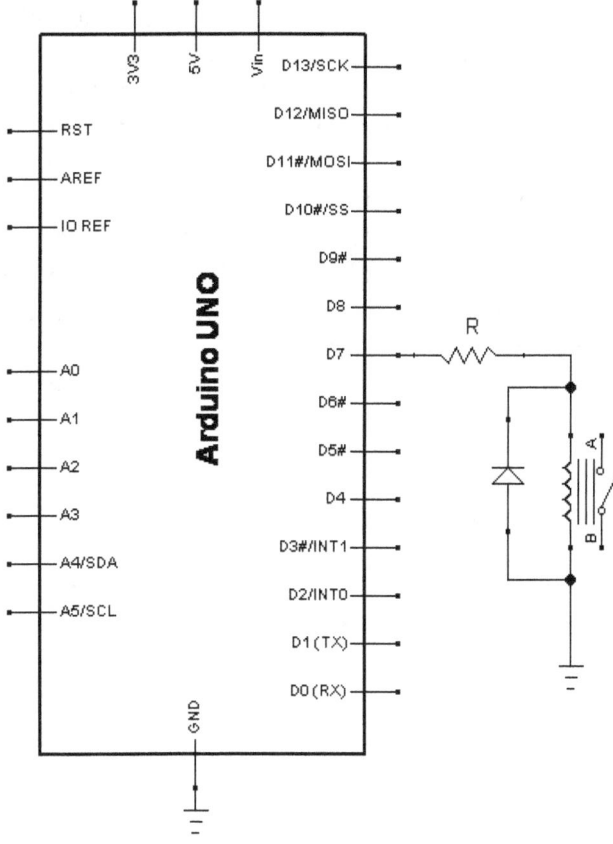

Figure 2.3

This circuit has the current limiting resistor discussed previously. The diode is for protection against a current spike when the power from the Arduino to the relay is switched suddenly from positive to ground. Note that as long as the voltage from the Arduino output is positive, no current will flow through the diode. It is only when the relay sends a surge that the diode conducts.

Normally you do not need to worry about the voltage on Arduino output pins going above the rated voltage, since most loads are passive. However, in later chapters (in particular, chapter 7) I will discuss connecting the Arduino output pins to circuits with external power supplies, and it is conceivable that a short circuit or other unusual circumstance could generate a positive voltage greater than the allowed voltage of the Arduino pins and feed this into the Arduino pin. In that case, you might want to use a zener diode on the output pins. You would, of course, use a zener diode with a threshold voltage greater than the normal output of your Arduino digital output pin, which is normally 5 volts. You do not want the zener diode to conduct current when the Arduino output goes HIGH.

You can also connect a diode to the output pin to avoid a reverse current, as shown in Figure 2.4.

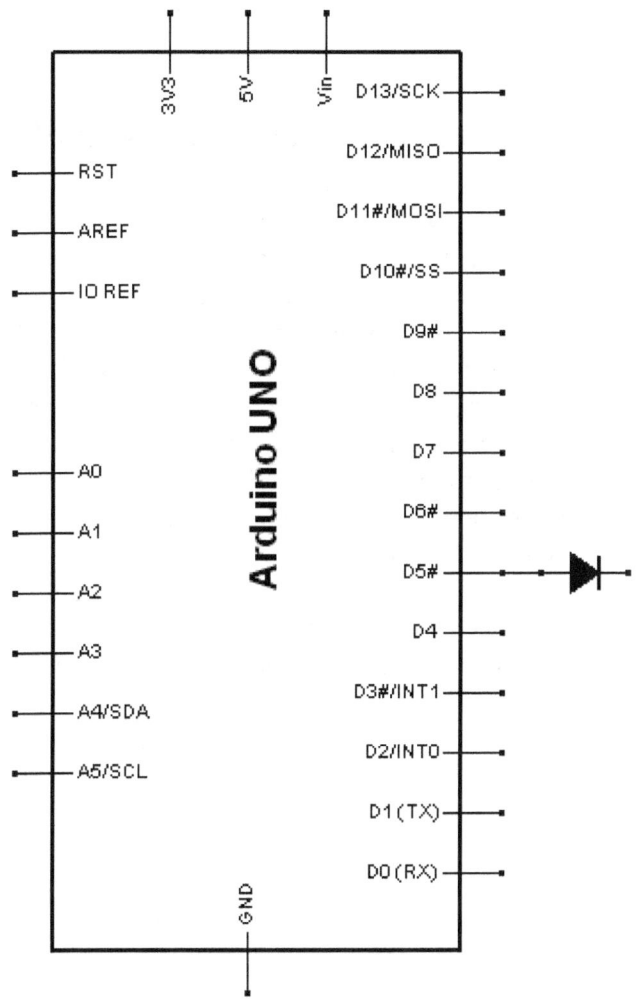

Figure 2.4

Chapter 3

Measuring resistance with Arduino

One of the most important measurements you can make with an Arduino is resistance. This might seem strange at first, since most people at all familiar with electronics associate resistance with the value of a resister. However, most types of sensors (thermistors, photo resistors, moisture detectors, etc.) are variable resistors that change resistance depending of temperature, light, or other environmental factors. Therefore, being able to measure resistance allows you to measure the readings of many types of sensors.

Since the Arduino analog inputs measure voltage, you need to translate the resistance into voltage. You do this with a voltage divider, shown in Figure 3.1.

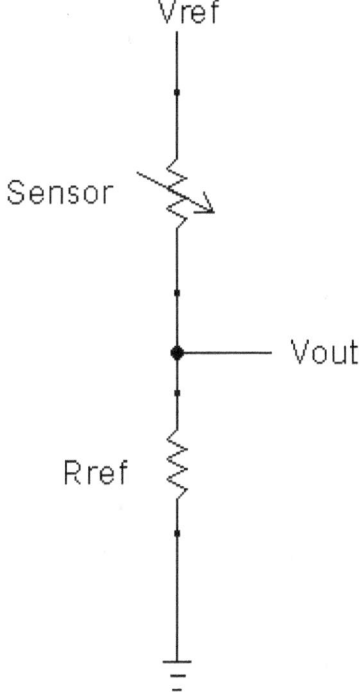

Figure 3.1

In this diagram, I have used a symbol for a variable resistor as a general symbol to represent any sensor whose resistance varies depending on the environment. For this circuit,

Vout = Vref * Rref/(Rref + Rs)

where Rs is the resistance of the sensor, Rref is the resistance of a reference resistor, and Vref is a reference voltage. Using a little algebra,

Rs = Rref *(Vref/Vout – 1).

For Vref, you want to use the 5 volt reference pin of the Arduino itself. Vout would go to any Arduino analog input, and the ground goes to the Arduino GND pin. The circuit is shown in Figure 3.2.

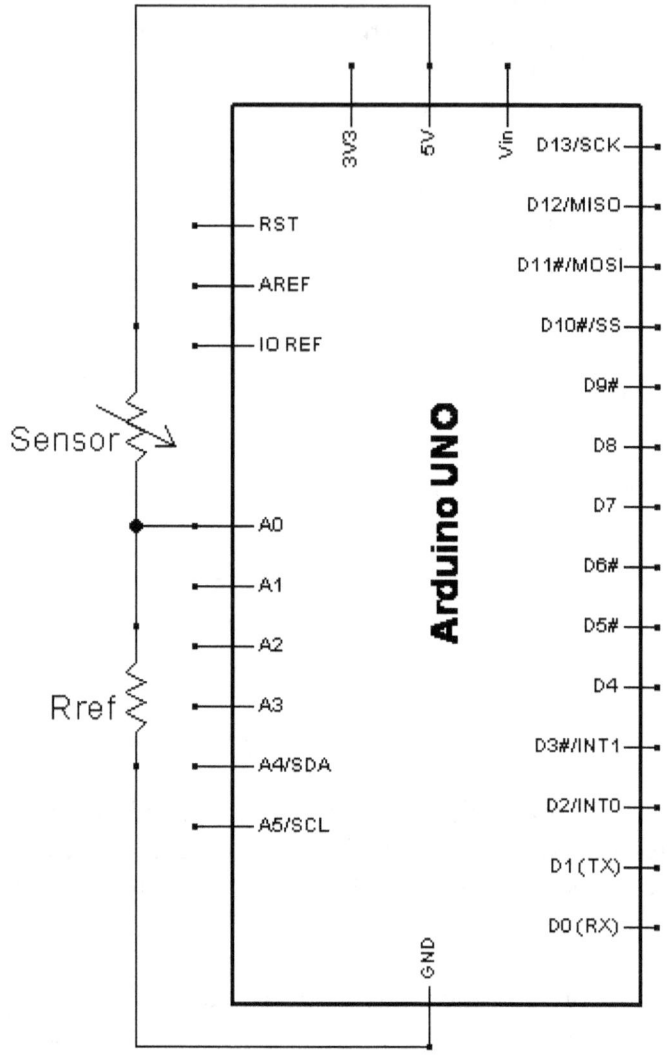

Figure 3.2

Since the reading of an analog pin is 204.6 times the voltage and Vref is 5, the equation
$$Rs = Rref *(Vref/Vout - 1)$$
becomes

Rs = Rref * (5/(A0 / 204.6) -1)
which becomes
Rs = Rref * (1023/A0 – 1)
where A0 is the numeric value returned by the analog pin
A0. The only value you need to supply for this equation is
Rref. Program this equation into your Arduino, and it will
give you the value of Rs. Use float variables for Rs and
Rref.

Remember the discussion in Chapter 1 about
accuracy of the resistor values. The accuracy of your Rref
determines the accuracy of your Rs reading. If Rref is, for
example, 10% more or less than you think it is, your
reading for Rs will be off by 10%. Try to use a high
precision resistor and/or check it with a high precision
resistance meter to make sure you are putting the right
value for Rref into your code.

This equation is nonlinear. Changes of Rs at lower
values of Rs give a greater increase in voltage at A0 than
changes of Rs when Rs is high. The amount of nonlinearity
will depend on the value of Rref compared to Rs. If Rref is
equal to the highest value that Rs takes, the change is more
linear, about 4 times as rapid if the value of Rs is near 0 as
it is if the value of Rs is near the maximum. If Rref is 10%
of the maximum value of Rs, Vout changes about 110
times as fast when Rs is near 0 as when it is near
maximum. As Rref gets smaller, the difference in Vout gets
smaller at high resistances of Rs and larger as Rs
approaches 0. The difference in voltage output near
maximum Rs is about 3 times as much when Rref equals
maximum Rs as it is if Rref equals 10% of maximum Rs.
Remember that the Arduino measures the voltage in integer
increments, so near the top range of Rs, the difference in
Vout might be too small for the Arduino to measure. I
therefore recommend using a reference resistor equal to the
highest value of Rs. Values of Rref above the maximum for
Rs decreases the voltage change at both lower and higher
values of Rs, so do not use an Rref greater than the
maximum for Rs. One other point to note is that the total

voltage change from maximum Rs to Rs equals 5 volts to 2.5 if Rref equals maximum Rref.

There is a circuit that gives you a linear relationship between the voltage to the analog Arduino input and the resistance. This means that you get a much greater change in the voltage at the high end of the resistance. This circuit also uses more of the input range of the Arduino input. It does, however, require a more complex circuit. This circuit is shown in Figure 3.3.

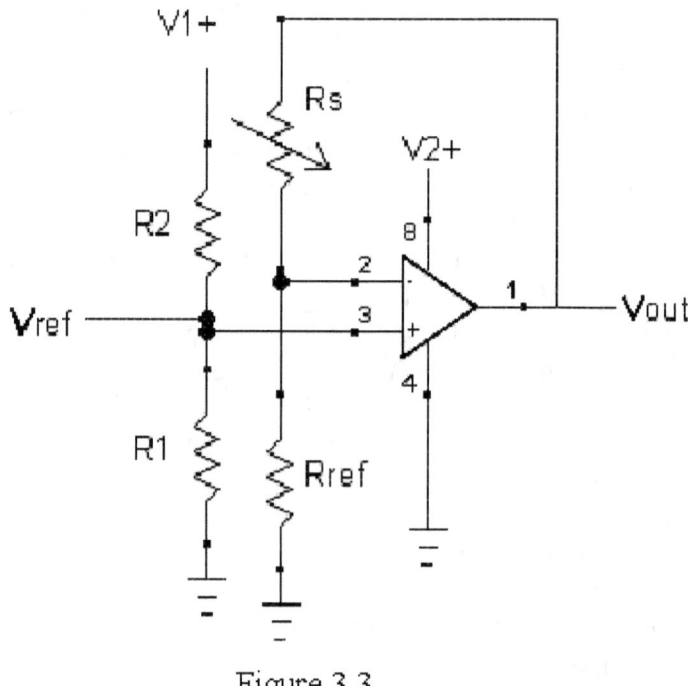

Figure 3.3

This circuit uses an operational amplifier. In this circuit,

Vout = Vref * (Rs/Rref +1)

where Vref is a reference voltage you set by selecting R2 and R1, Rs is the sensor resistor, and Rref is the reference resistor. Applying some algebra

Rs = Rref * (Vout/Vref – 1)

You will note on the diagram that there are two source voltages listed, V1+ and V2+. This is because the voltage for the voltage divider does not have to be the same as the power supply voltage for the op-amp. V1 should be 5 volts off the Arduino 5V pin. V2 can also be this 5 volt supply. However, there is a small problem with the op-amp output. On most op-amps, the output cannot go above the power supply minus about 1.5 volts. Therefore, if V2 is 5 volts, Vout cannot go above about 3.5 volts. If the Arduino analog reference is the default value of 5 volts, your analog pin will to provide a reading above 718 instead of being able to go to 1023, so you will lose resolution in your readings.

There is a very simple solution to this. You can connect a voltage between 0 and 5 volts to the AREF pin to set the high end of the analog readings. For example, connecting the 3.3 volt connection of your Arduino to the AREF connection will set the 1023 top of the analog reading to equal 3.3 volts instead of 5 volts. This means that the number you read from the analog pins will be 3.3/1023 instead of 5/1023, giving you greater precision. Instead of connecting AREF to the 3.3 volt Arduino pin, you can also use a voltage regulator to pick whatever voltage you like to your AREF pin. Another possibility is to use a zener diode and resistor to select a reference voltage. This is shown in Figure 3.4

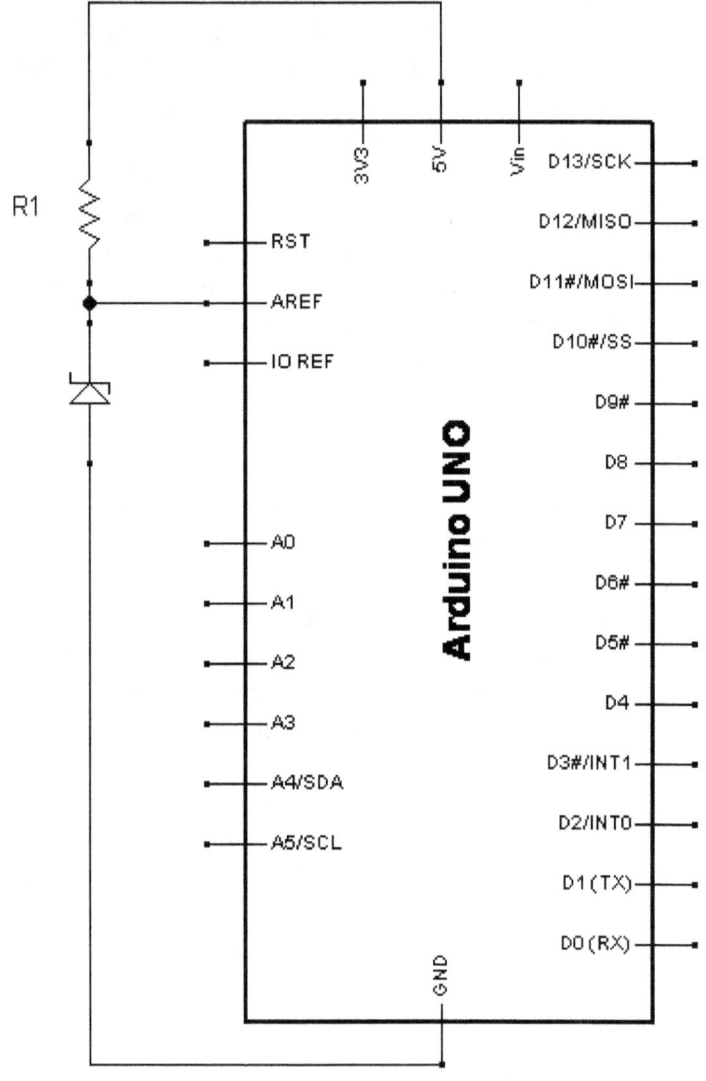

Figure 3.4

R1 should be about 100 ohms. There are various commonly available values for the zener diode. A good one for this purpose is 3.9 V. That sets the top of the analog input range just above the limit of what the op-amp output.

Note: If you do connect a reference voltage to the AREF pin, you MUST include the line analogReference(EXTERNAL); in your void setup section. Failure to do this can damage your Arduino when you connect the reference voltage to the AREF pin. In addition, you must be sure to select the values of Rref, R1 and R2 so that Vout cannot exceed the reference volts, such as 3.3 volts. This will probably not be a problem is you use a 3.9 V zener to set your reference voltage.

If your circuit is powered by a greater voltage, such as a 9 volt battery, powering your op-amp off this can be an advantage. There is one disadvantage to powering the op-amp at more than 6.5 volts: You will need to take care to set your R1, R2, and Rref values so that Vout cannot exceed 5 volts, or you might damage your Arduino.

As we saw previously in dealing with voltage dividers,

Vref = V1 * R1/(R1 + R2)

V1 will be 5 volts from your Arduino 5V pin. So if, for example, R1 is 680 ohms and R2 is 10 K ohms, Vref will be about 0.32 volts. If Rref is 11% of the highest value that Rs can be, Vout ranges from Vref (if Rs equals 0) to 3.23 volts, which is perfect if the op-amp is powered by 5 volts and you set AREF to 3.3 volts. If you want to leave some additional safety margin to make sure that Vout does not exceed 3.3 volts, you can use a slightly larger Rref.

If you power the op-amp with more than 6.5 volts so that Vout can reach 5 volts, you can make Rref 7% of the maximum value of Rs, which will allow Vout to reach about 4.7 volts. This leaves you a small margin of error to make sure Vout does not exceed the 5 volt Arduino analog input limit. This circuit is shown in Figure 3.5.

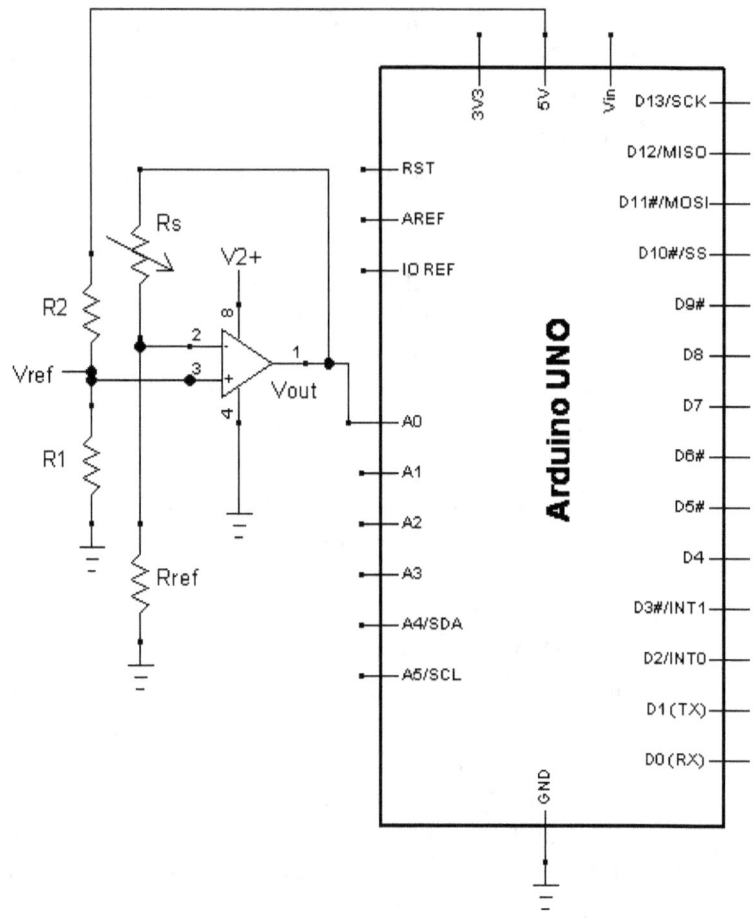

Figure 3.5

Since the reading on an Arduino analog input pin is 204.6 times the voltage,

Rs = Rref * ((A0 / 204.6)/Vref – 1)

The accuracy of your measurement will depend on your accuracy of Rref and Vref, and the accuracy of Vref depends on either the accuracy of R1 and R2, or of a measurement you make if you actually measure Vref and use that measurement in the equation. This is one adjustment to the circuit that you can make to reduce this error. The revised circuit is shown in Figure 3.6.

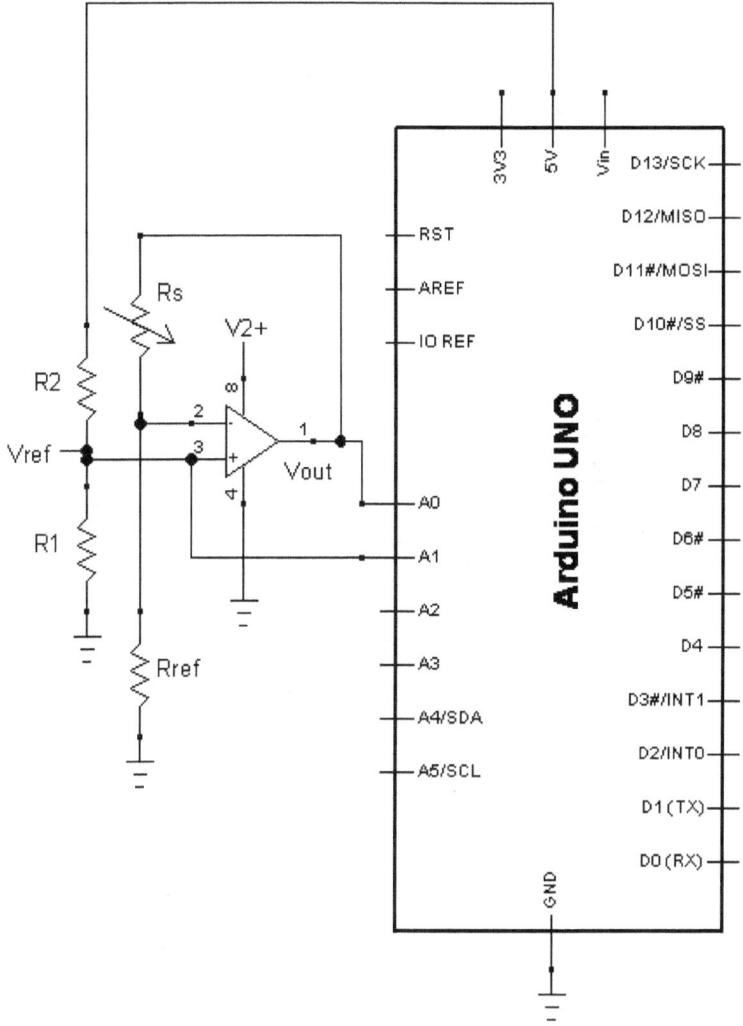

Figure 3.6

Here you are using the Arduino to measure Vref directly. The equation for Rs then becomes simplified to
Rs = Rref * (A0/A1 – 1)
You do not have to worry about the accuracy of your R1 and R2 resistors or accurately measuring Vref.

There is one concern. If Rs were to go to infinity, which might happen if Rs were disconnected or malfunctioned, Vout would go to V2 minus about 1.5. This

29

is no problem if V2 is 5 volts, but if you use a power supply above 6.5 volts, the voltage at A0 would go above 5 volts. To protect the Arduino from this, you can put a zener diode from A0 to ground and add R3 to provide the voltage drop if the zener diode kicks in, as shown in Figure 3.7.

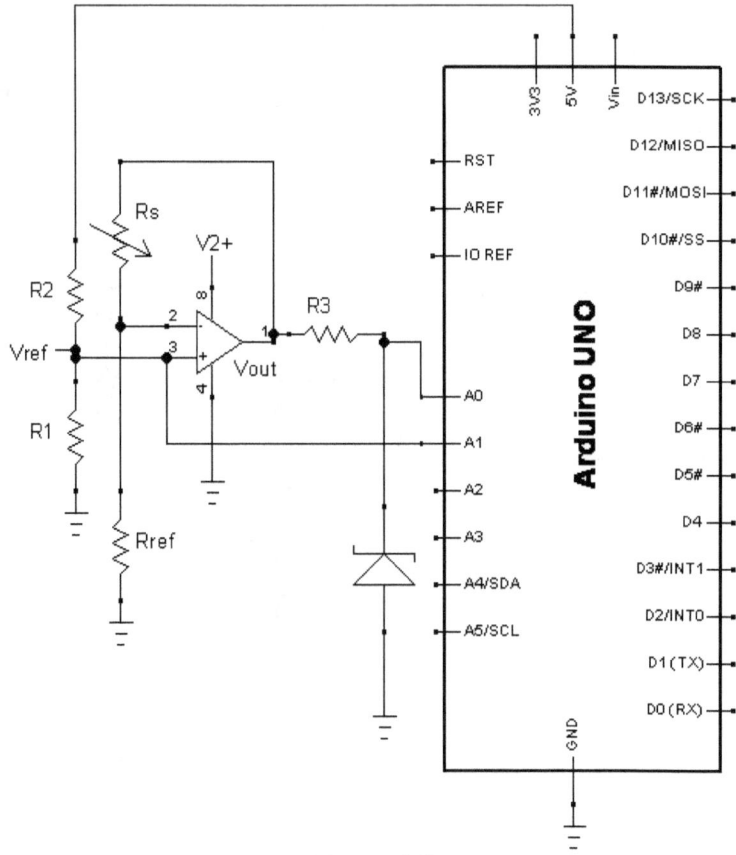

Figure 3.7

Using a 5.1 volt zener diode, this will prevent the voltage to A0 from exceeding 5.1 volts. Note that the reading of the voltage at A0 will be inaccurate, because the voltage will be limited to 5.1 volts regardless of what Vout the op amp is trying to produce, but that should not matter because the zener diode will only take effect if the circuit is malfunctioning, not during the normal range of Rs. The point is, do not use a zener diode to routinely limit the voltage to an analog input, only as a safety measure to protect against malfunction.

Chapter 4

Measuring DC voltages

Next to measuring resistance, measuring voltage is probably the most useful thing you can do with the analog inputs on an Arduino. Many sensors do output an actual voltage, such as thermocouples, photovoltaic cells, and microphones. There are also many reasons to want to measure voltage, such as monitoring a battery level.

Measuring voltage with an analog input should be simple. After all, that is exactly what it is designed to do. However, you do need to reduce the voltage if the voltage is outside the allowed range (0 to 5 volts). You may also need to amplify the voltage if it is very small. Ideally, the voltage should range from 0 to 5 volts, so you get the maximum resolution for your measurements. If, for example, the voltage only ranges from 0 to .1 volts, the reading on the analog input will only range from 0 to 20, which does not give much resolution.

Reducing the voltage if it exceeds 5 volts is simple. You just use a voltage divider. Unlike the circuit in Chapter 2, where the top of the voltage divider was connected to 5V on the Arduino, you connect the top of the voltage divider to the voltage source, as shown in Figure 4.1, where the positive and negative of the voltage you want to measure is connected to V+ and V-.

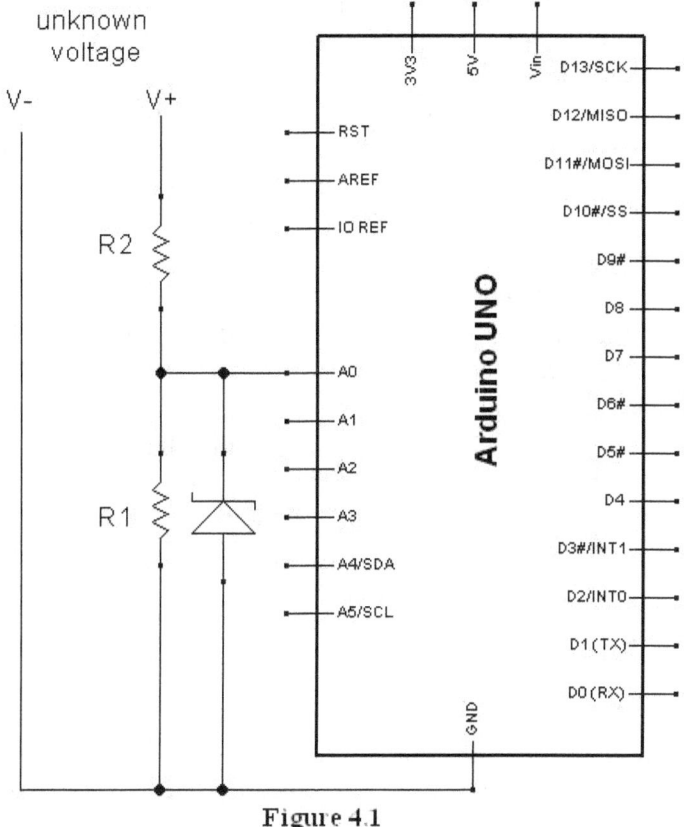

Figure 4.1

Using the standard formula for a voltage divider, the voltage at A0, Va, will be

Va = V * R1 /(R2 + R1)

Solving for R2 gives

R2 = R1 * (V/Va – 1)

You want Va to be a MAXIMUM of 5 volts, so R2 must be a MINIMUM of

R2 = R1 * (V/5 -1)

Where V is the MAXIMUM you expect the voltage you are measuring to ever reach. For example, if the maximum voltage you expect to measure is 100 volts and you select a 10,000 ohm resistor for R1, then the minimum value for R2 would be 10000*(100/5-1), which is 190,000 ohms. The closest common resistor to that would be 200 K.

This would mean that Va could reach 100*10000/(10000+200000), which is 4.762 volts. Note that you want to use fairly large resistors so that you do not draw much current from the voltage you are testing. Using a 10K and a 200K resistor for R1 and R2 will draw about 0.5 milliamps, which is quite satisfactory.

Solving for V to find out what that is (since that is what you are trying to measure),

$V = Va * (R1 + R2)/R1$

Since $Va = A0 / 204.6$ (where A0 is the numeric value that you read from A0),

$V = A0 * (R1 + R2)/(R1 * 204.6)$

This is good for reducing voltage if the voltage is too high. If the voltage is too low, you can use an op amp to magnify the voltage. The circuit is shown in Figure 4.2.

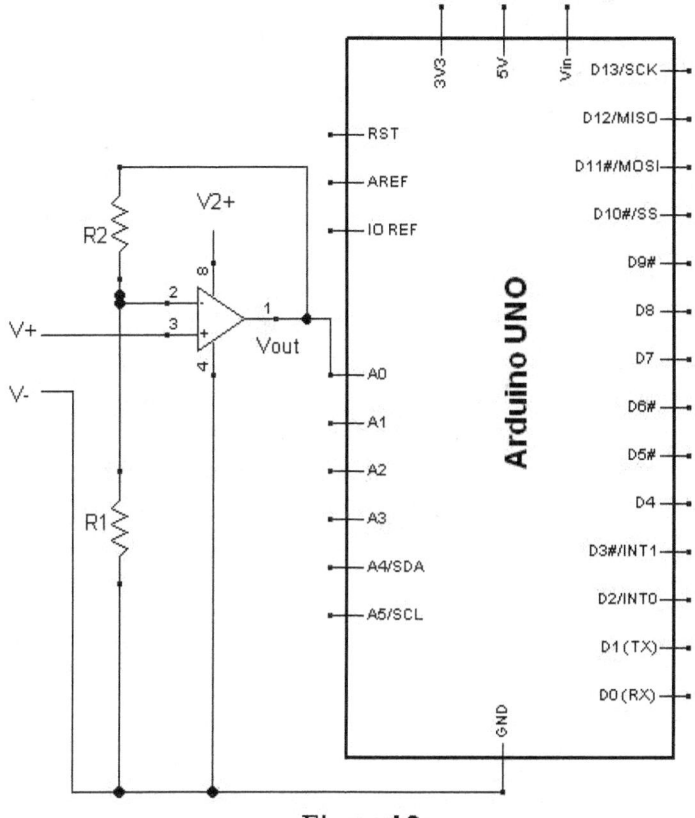

Figure 4.2

The input of the circuit is V+ and V-, where you connect the positive and negative of the voltage you want to measure. The output of the op amp is Vout.

Vout = V*(R2/R1 + 1)

For example, if R2 is 10K ohms and R1 is 1K ohms, Vout will be 11 times V.

You want to select R1 and R2 carefully. If the op amp power (V2) is 5 volts, the maximum Vout will be 3.5 volts (due to the limitation on the op amp), so you so need to make sure that the maximum V you measure times (R2/R1 + 1) is not more than 3.5, or you will not get accurate readings when V exceeds that level. That is, once V reaches a level such that V*(R2/R1+1) exceeds 3.5, you will get the same reading from the Arduino regardless of what V is. If you power the op amp with more than about 6.5 volts, you want to make sure that Vout cannot reach more than 5 volts, because of the limit on the voltage input of an Arduino analog input. In calculating the value to use for R2,

R2=R1*(Vout/V – 1)

where V is the maximum value that you expect the voltage you are testing to reach, and Vout is the maximum voltage you want for the op amp output (3.5 volts or 5 volts, depending on the op amp power, V2). This equation gives you the MAXIMUM value you can use for R2.

If you do power the op amp with more than 6.5 volts, you might want to add a zener diode and resistor to protect the Arduino, in case V ever goes higher than expected and drives Vout above 5 volts. This is shown in Figure 4.3.

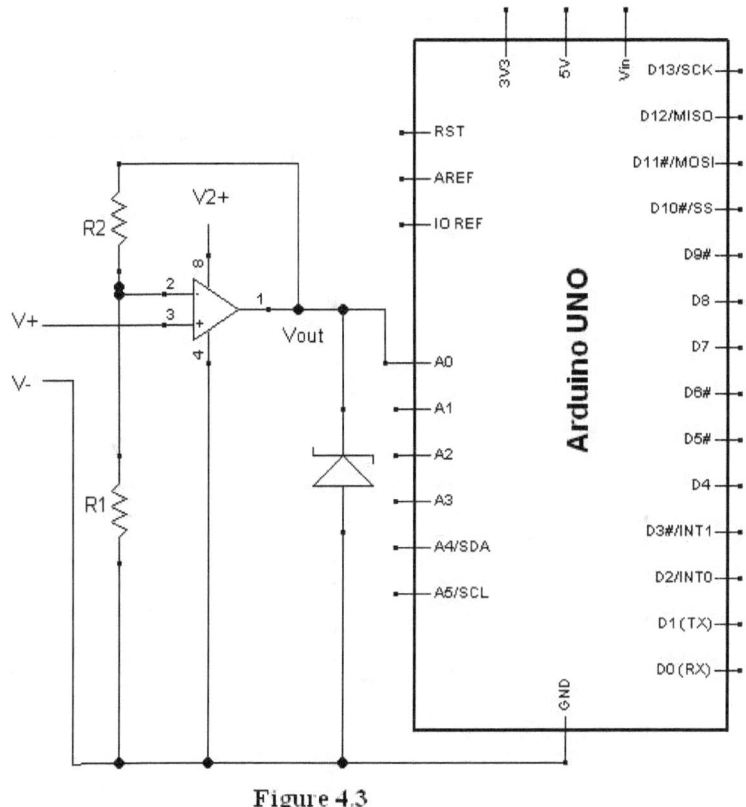

Figure 4.3

One potentially very practical use for measuring voltage is to keep track of the voltage of your Arduino's power supply. For example, if you are powering your Arduino with a 9 V battery, you might want to know when the battery is getting low. The schematic for this is shown in Figure 4.4.

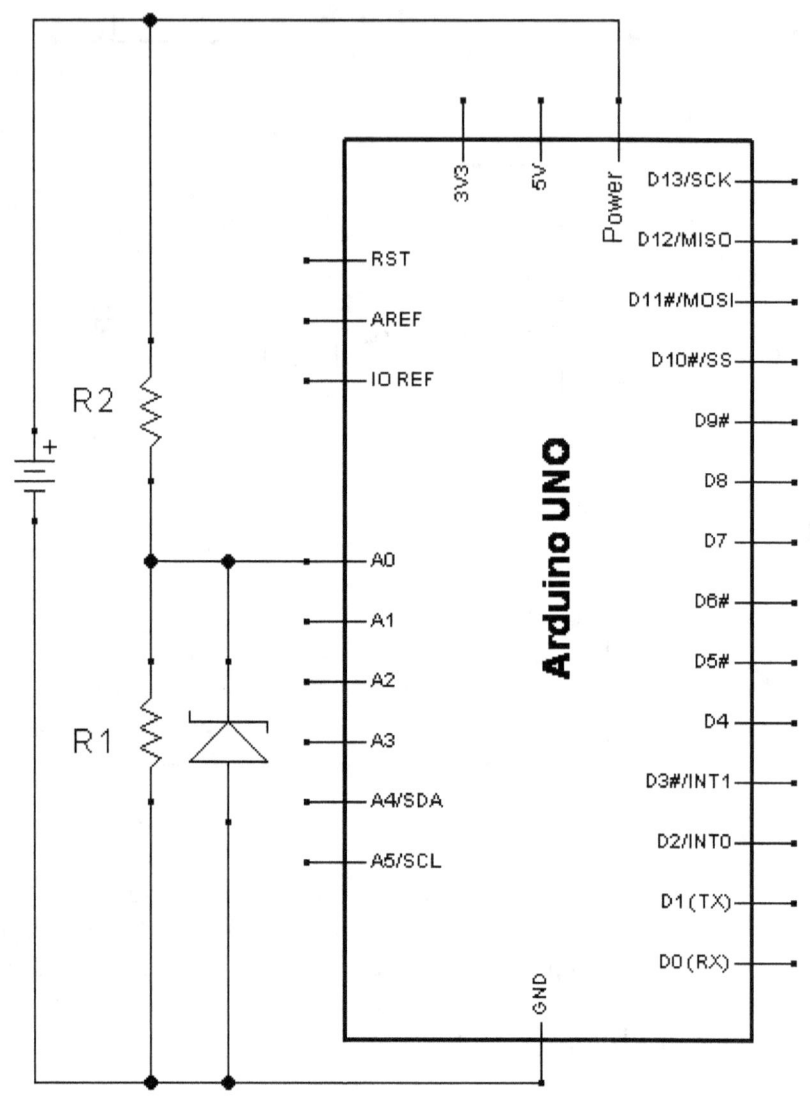

Figure 4.4

Chapter 5

Measuring AC voltages

Although it is not as common as measuring DC voltage or resistances, you might sometimes want to measure alternating currents (AC). For example, you might want to monitor an AC power supply coming from a transformer or even a wall outlet. Another example would be the output of a microphone or some other vibration monitoring equipment. Figure 5.1 shows an AC sine wave signal (the most common AC wave form) on an oscilloscope.

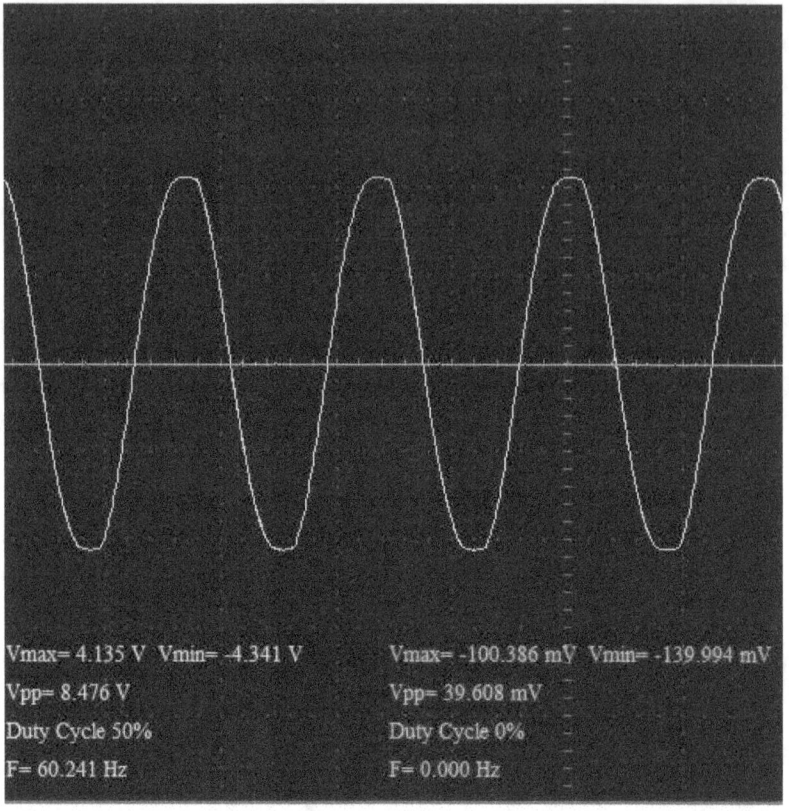

Vmax= 4.135 V Vmin= -4.341 V	Vmax= -100.386 mV Vmin= -139.994 mV
Vpp= 8.476 V	Vpp= 39.608 mV
Duty Cycle 50%	Duty Cycle 0%
F= 60.241 Hz	F= 0.000 Hz

Figure 5.1

There are several problems with monitoring AC. For one thing, it goes negative for part of its cycle, and you should definitely not apply a negative current to an Arduino analog (or any other) pin. You need a way to only have the positive part of the signal. Second, since the signal is constantly changing throughout the cycle, you need to sample or stabilize it.

The circuit shown in Figure 5.2 solves the problem of the negative voltage.

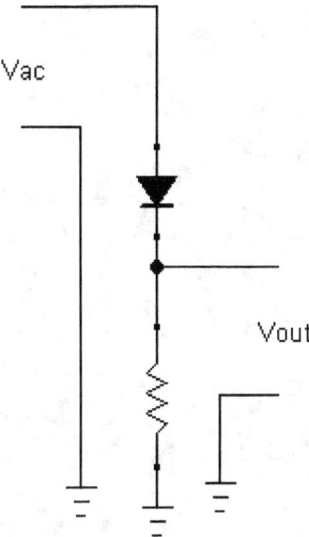

Figure 5.2

The diode blocks the negative current, resulting in the wave form shown in Figure 5.3.

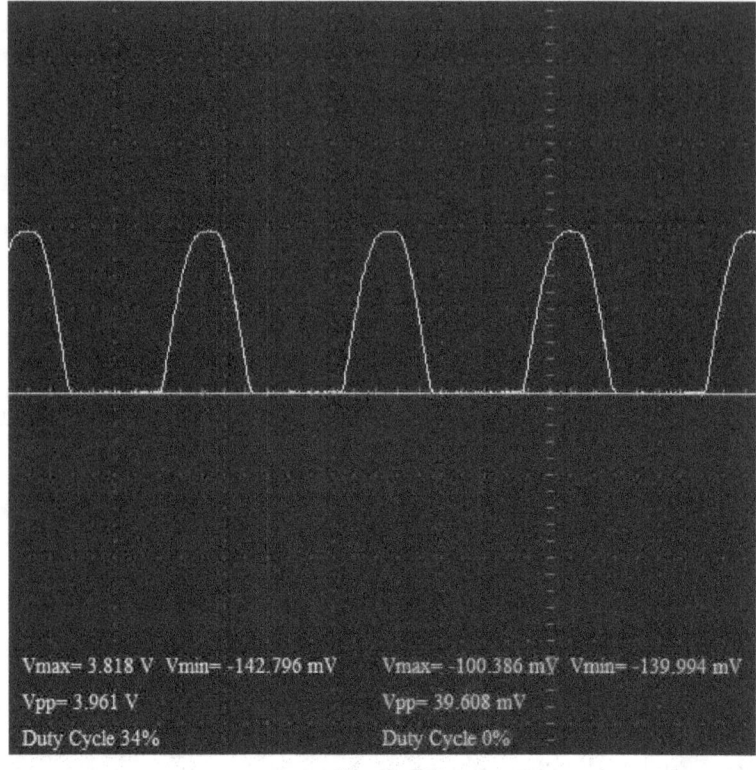

Vmax= 3.818 V Vmin= -142.796 mV   Vmax= -100.386 mV Vmin= -139.994 mV

Vpp= 3.961 V   Vpp= 39.608 mV

Duty Cycle 34%   Duty Cycle 0%

Figure 5.3

You can capture the high point of the voltage (or at least, close to it) by replacing the resistor with a capacitor, as shown in Figure 5.4.

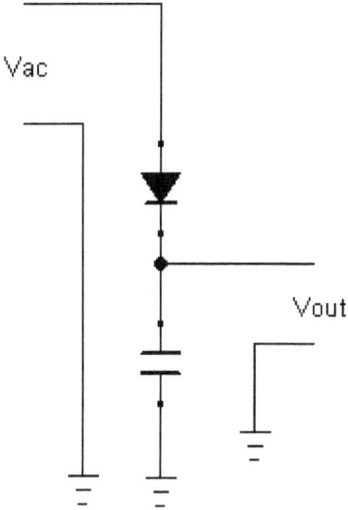

Vac

Vout

Figure 5.4

The current from the diode will charge the capacitor up to nearly the peak positive voltage of the AC wave form. It does not quite reach the peak, because to voltage drop across the diode, and some leakage across the capacitor and also whatever you connect to Vout to measure it (in this case, the Arduino).

One minor problem with this circuit is that it does not provide a way to discharge the capacitor. Of course, a real capacitor will discharge due to leakage within the capacitor and some small current flow out Vout, but if you want to measure a variable AC voltage, you might want to add a large resistance resistor and shown in Figure 5.5.

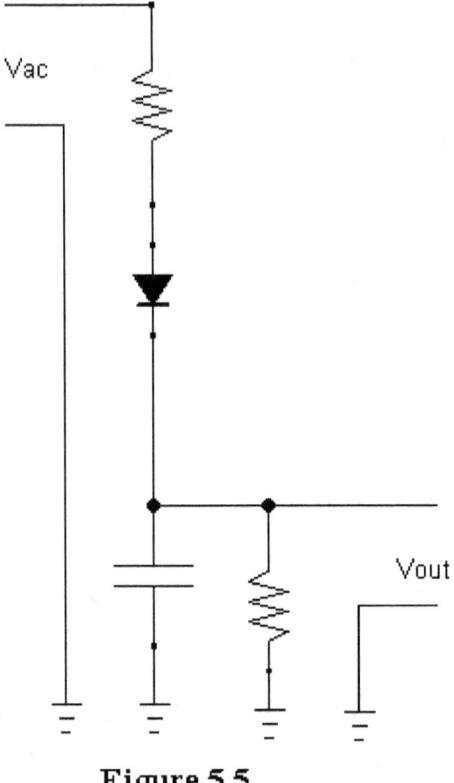

Figure 5.5

     The circuit in Figure 5.6 shows this circuit connected to the Arduino.

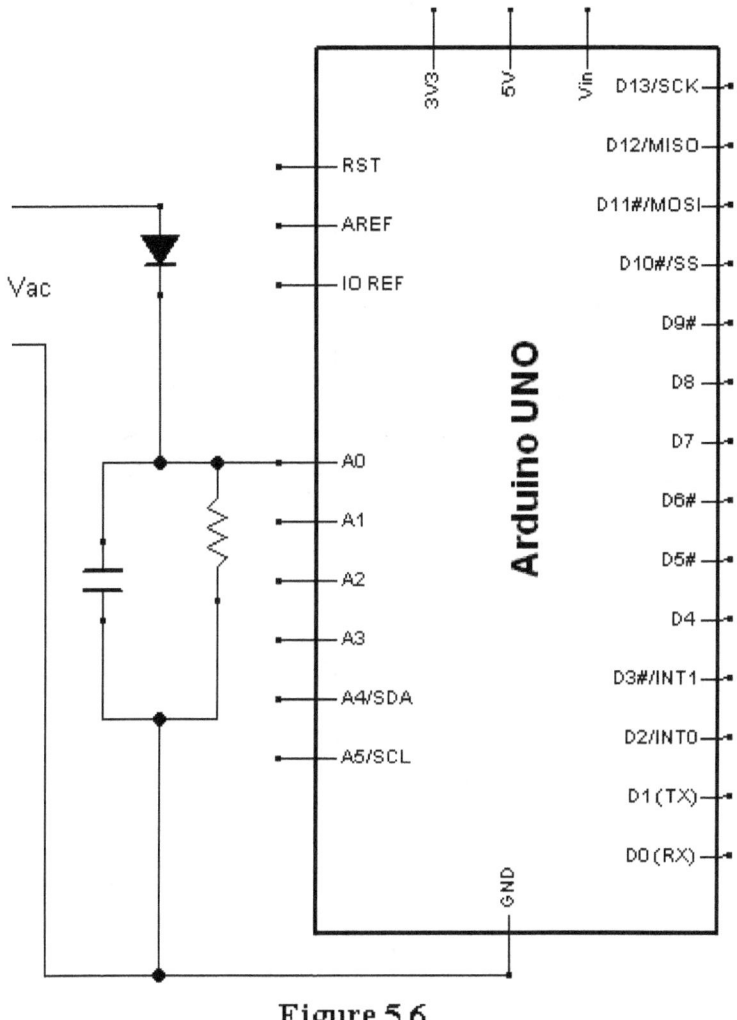

Figure 5.6

This circuit works as long as the voltage you want to measure does not exceed 5 volts. If the voltage can exceed this, you need to take the connection to the Arduino analog input from a voltage divider, as shown in figure 5.7.

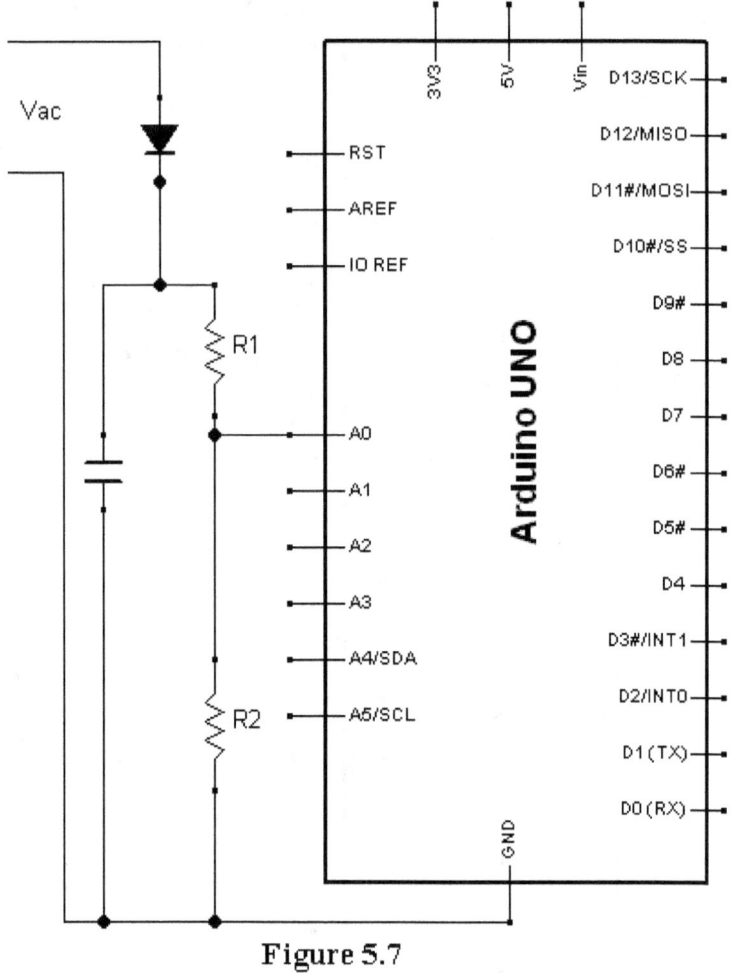

Figure 5.7

One small problem with this circuit is that the capacitor must be rated at the full peak voltage of Vac. In addition, the rectifier must pass considerable current as it charges the capacitor, and also must stop the twice the peak current of Vac, because when the capacitor is fully charged and Vac goes negative, it is resisting the charge on the capacitor trying to flow into the reverse voltage of Vac. It might be hard to find capacitors and diodes that can handle higher voltages, like 120 volts even or 220 volts. A solution to this is to put the voltage divider BEFORE the diode and capacitor, instead of after it, and shown in Figure 5.8.

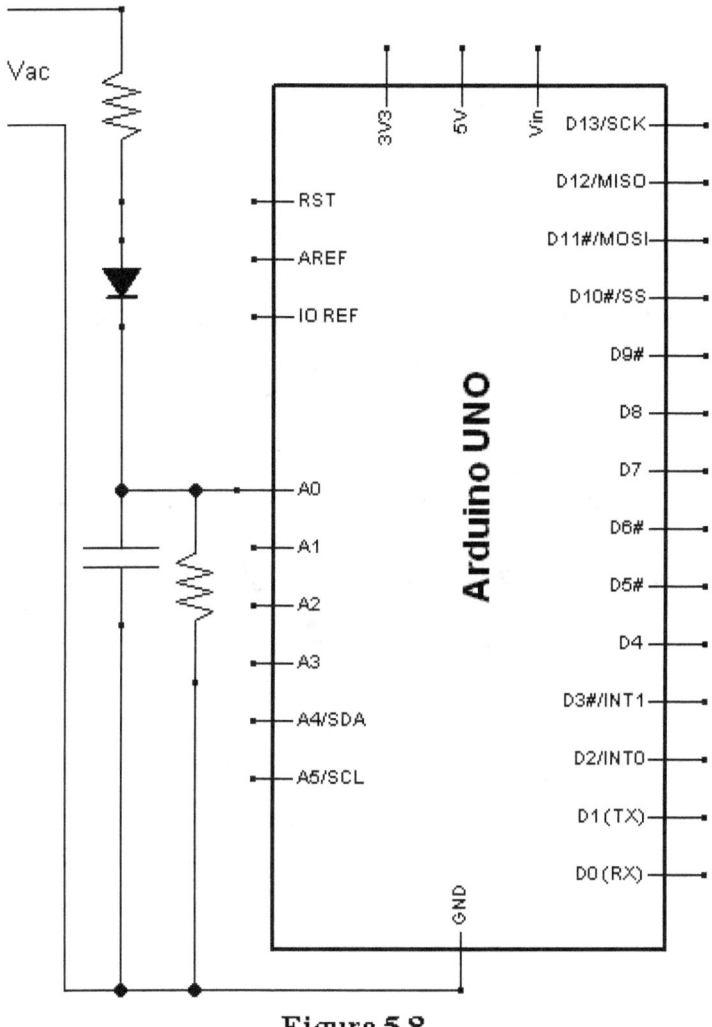

Figure 5.8

Chapter 6

Adjusting trigger voltages on digital inputs

As mentioned in Chapter 1, a digital input pin will register HIGH if the input voltage is over 3 volts and LOW if it is under 2 volts. Suppose have an output from a sensor or other device that is lower than 3 volts. For example, suppose you have an output that goes from 0 to 1 volt. You could use an analog input, but there are some disadvantages to these. For one thing, the response time on analog inputs is about 100 microseconds, so the maximum sampling rate is about 10,000 times per second. For another, there are fewer analog input pins than digital pins on an Arduino, so you cannot sample as many things. Therefore, you might want to be able to trigger the digital input pin to go HIGH at some lower or higher voltage.

One way to do this is with an op-amp. The circuit is shown in Figure 6.1.

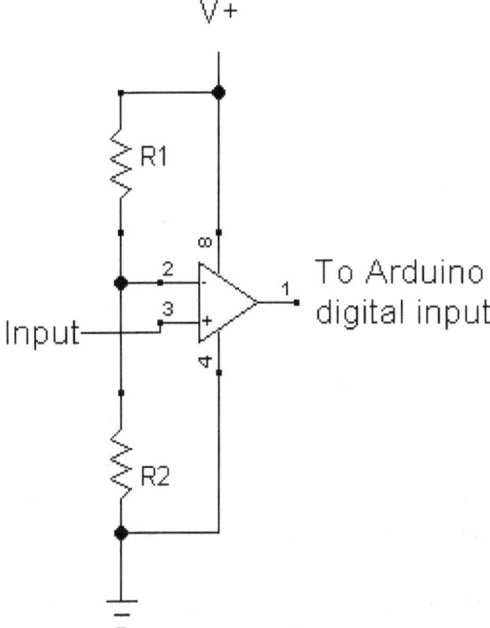

Figure 6.1

If the voltage at the non-inverting input is higher than the voltage at the inverting input, the output of the op-amp will go to maximum. Assuming the op-amp is powered by the 5 volt Vin pin of the Arduino, this would be about 3.5 volts. This is enough to cause the digital input to clearly read as HIGH. If the input at the non-inverting input is lower than the inverting input, the output voltage will go to minimum (normally about 0 volts). Since you can control the voltage at the inverting input of the op-amp, you can control the trigger voltage for the output.

Chapter 7

Controlling high current or high voltage devices with Arduino

The most common way to control high currents or voltages over 5 V with an Arduino is to use relays. However, there are some advantages to using transistors. For one thing, they are cheaper. For another, they have faster response times. This can be important if you want to create a dimmer circuit. With the PWM outputs of an Arduino, you can vary the effective voltage of your output by rapidly switching the power on and off. You can use this, for example, to dim an LED. However, if you want to dim a higher power device, like a 110 V light, you cannot connect it directly to the Arduino output, which is only 5 volts. You cannot use a relay, because a relay cannot open and close fast enough. There are other advantages to transistors over relays, such as being quieter, drawing less power to operate, and lasting longer with frequent switching.

Transistors simply have an input connection where you apply a control voltage and another connection where you connect a load to. Then there is the ground connection, which applies to both the input and the load. There are various types of transistors. The most useful for us might be the bipolar junction transistor (BJT) and the metal–oxide–semiconductor field-effect transistor (MOSFET). The most useful type for connecting to the Arduino will be the MOSFET, because these have very high resistance at the gate, which you will be connecting the Arduino output to, so they draw almost no current. They also have very little resistance at the drain when they are conducting, so they make good switches. We will be using transistors only as switches with the Arduino, since the output of the Arduino is digital.

Figure 7.1 shows a MOSFET connected to an Arduino.

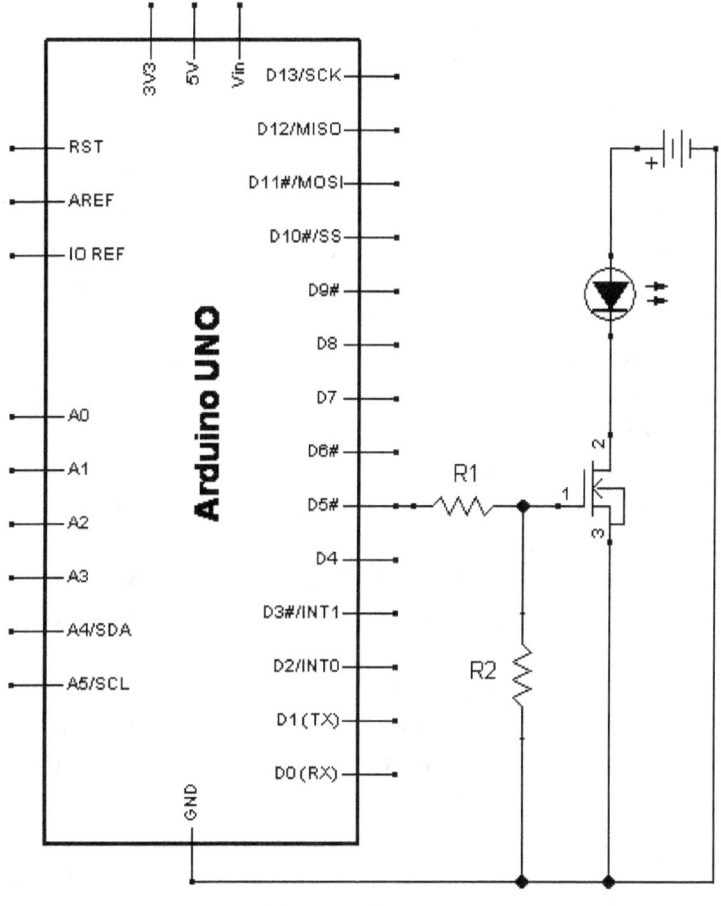

Figure 7.1

I have arbitrarily connected the transistor to output D5, but it could be connected to any of the digital outputs. The R1 is just a safety to make sure that too much current is not drawn from the Arduino in case of a short or failure of the MOSFET, but normally there will be almost no current to the MOSFET gate and no voltage drop across the resistor. At typical value might be something like 220 ohms. R2 has a more important role. The gate of a MOSFET actually has some capacitance, so it can actually hold a charge. If you apply a positive voltage to the gate and then disconnect, the gate can actually hold the charge for a long time, causing the MOSFET to remain on. Normally this is not a problem with the Arduino, because the output is either HIGH or LOW, and when it goes LOW it will discharge the MOSFET gate. However, if power to the Arduino is turned off while the output is HIGH, the gate can remain charged and the MOSFET continue to conduct. In addition, the gate is so sensitive that even radio waves or static electricity can charge the gate if the Arduino is disconnected or turned off. I recommend including R2 so that the gate is pulled to ground if no ongoing positive voltage is applied to the gate, to make OFF the default state for the MOSFET. A fairly high resistance, like 10 K or even 100 K is sufficient to discharge the MOSFET and will not significantly affect the circuit. The actual values are not really that important, so long as R2 is many times as high as R1, so that you do not have a significant voltage drop across R1.

I have used an LED to represent the load, but it could be any load. I have drawn a battery to power the load to dramatize the fact that the power supply of the load can be different than the power supply of the Arduino. In fact, it very often will be, since the whole purpose of the transistor is to allow the Arduino to control something that needs a much higher voltage than the Arduino. However, if you are powering something that runs off the same voltage as the Arduino but draws too much current for an Arduino digital output to provide, you could use the same power source for the Arduino as the load.

There are various MOSFETs you can use. One popular model for Arduino circuits is the IRF540, which you can buy on eBay for about $0.55 each when you buy them in batches of 10 if you do not mind waiting for a shipment from China. Figure 7.2 shows what the IRF540 looks like. In fact, many MOSFETs look like this. It is a common configuration.

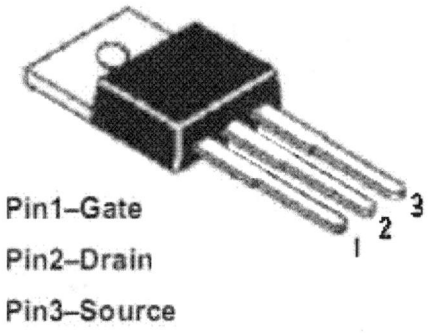

Pin1–Gate
Pin2–Drain
Pin3–Source

Figure 7.2

The IRF540 has a maximum drain to source voltage of 100 V, a maximum current of 33 amps at room temperature (maximum current allowed drops to 23 amps as the temperature rises to 100 C in case you plan on operating it in boiling water), and takes between 2 and 4 volts at the gate to cause it to pass a current (called the threshold voltage). The metal backing is a heat sink, and you can bolt it to a larger heat sink. If you need to use a voltage greater than 100 V in your circuit, there is the 2SK3568, which has a maximum drain to source voltage of 500 V, a maximum current of 12 amps, and also threshold voltage of 2 to 4 volts. There are many more types of MOSFET. The only real requirement of the MOSFET is that it has a threshold voltage less than 5 V (preferably closer to 3 V) so that the Arduino digital output of 5 V can trigger it. I do prefer to use N channel MOSFETs, which are the ones that the circuits in this book are designed for, because the voltage from the Arduino goes to ground, not the load.

If you need to draw more current than your MOSFET can handle, you can actually connect more

MOSFETs in parallel. Just connect the gates of all the MOSFETs together, the drains of all the MOSFETs together, and the sources of all the MOSFETs together. This is shown in Figure 7.3.

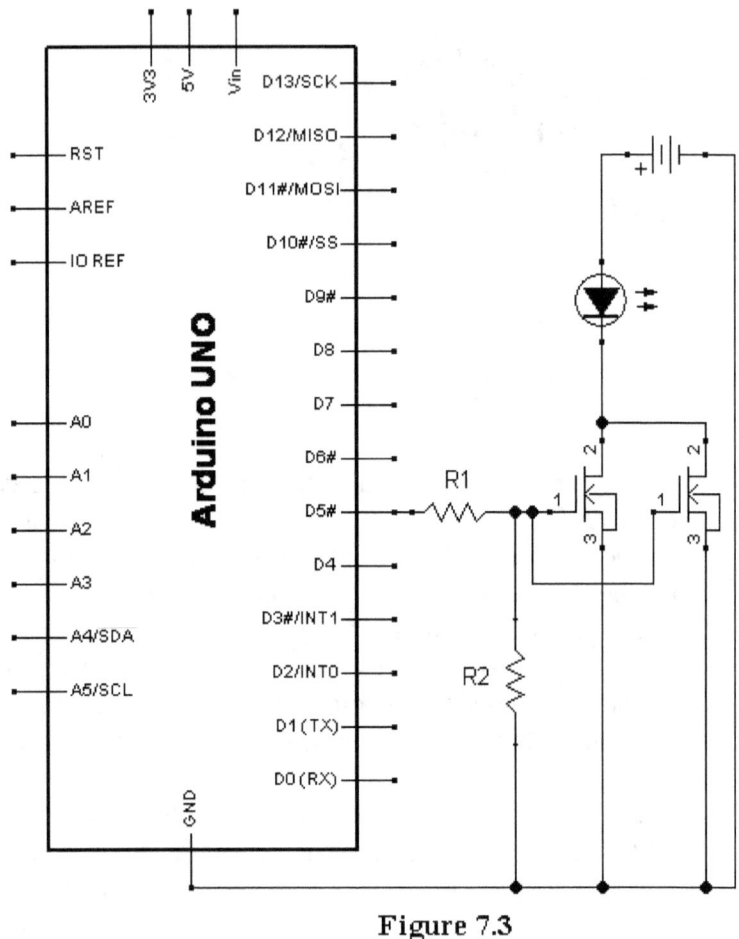

Figure 7.3

Figure 7.4 shows an oscilloscope trace of the voltage output from the Arduino bottom trace) and the voltage across the MOSFET (top trace). The output is pulse width modulated. This demonstrates the practicality of using a MOSFET to control a large load. Notice that the Arduino output is directly the opposite of the voltage across the MOSFET. This is because when the output of the

Arduino is HIGH, the MOSFET conducts and the voltage across it drops, allowing the voltage across the load to go up.

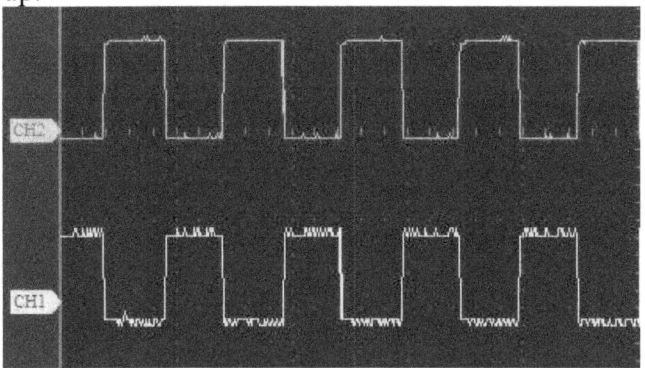

Figure 7.4

You can use BJT transistors instead of MOSFET if you want to. I recommend using NPN transistors if you do, because the load goes on the collector and does not interfere with the base to emitter current flow. An Arduino circuit using a BJT is shown in Figure 7.5.

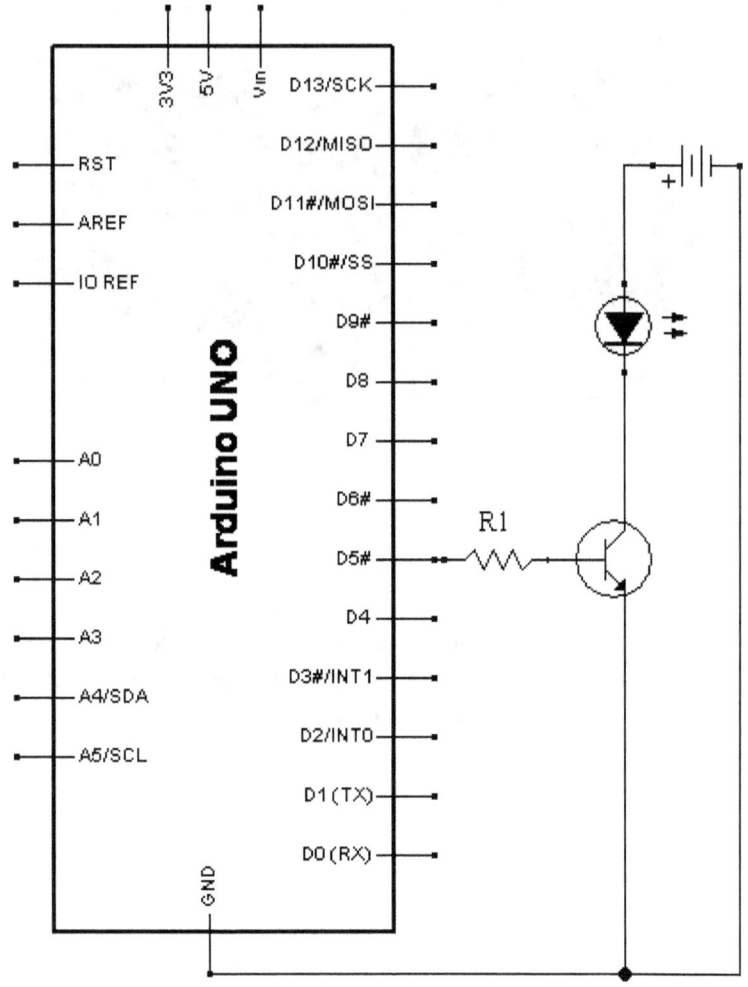

Figure 7.5

R1 is a current limiting resistor, to prevent the transistor from drawing too much current. The exact value you use can depend somewhat on which transistor you use and the exact value is not generally important. You want to use the highest resistance that will pass enough current to saturate the transistor base and allow the transistor to conduct enough current for your load. Generally, a 2.2K resistor is about right. If that is not allowing enough current to your load, reduce it to about 1 K.

As for what type of transistor to use, the TIP31C is a usually a good one. It can conduct 3 amps and can handle up to 100 V from the collector to the emitter, meaning that it can operate up to 100 V loads. If you need to operate at higher voltages, the MPSA42 can handle 200 V from the collector to the base. It can only handle about .5 amps (half an amp), however. If you need to conduct more, you can string them in parallel, as shown in Figure 7.6.

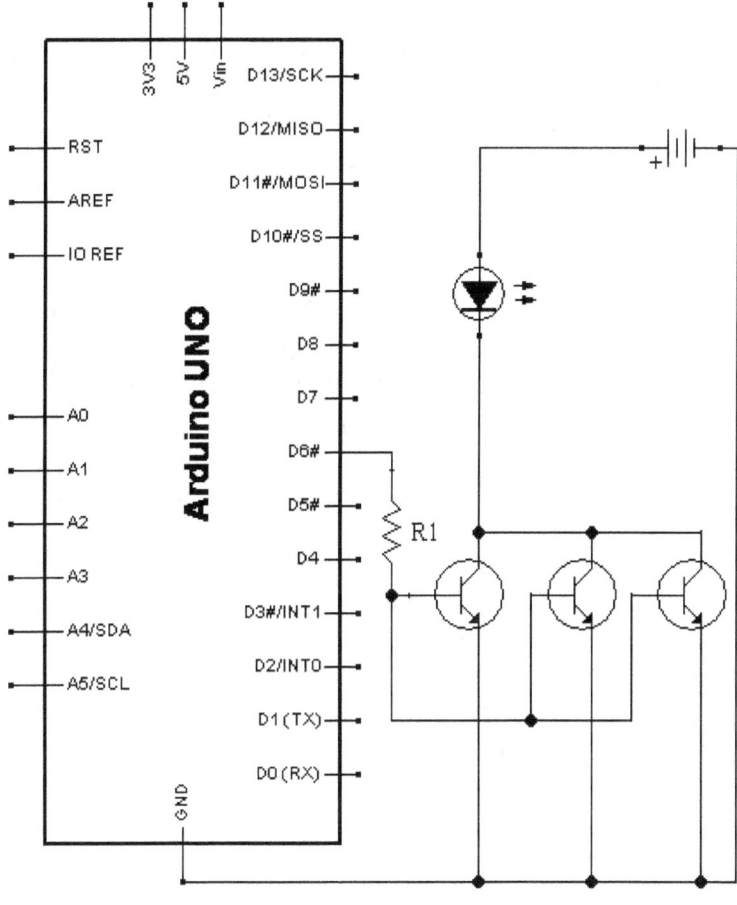

Figure 7.6

Chapter 8

True Analog output. Converting PWM to analog.

The output the digital outputs is always either 0 or 5 volts. In order to provide an apparent variable voltage output, some of the output pins can rapidly flip back and forth from 0 to 5 V, at a rate of about 980 times per second. It is not necessarily at 5 V half the time. You can control it from 0% of the time to 100% of the time using the analogWrite(pin,value) function. If value is 0, the output will be 0 V all the time. If value is 255, it will be 5 all the time. If, for example, value is 122, it will be 5 V half the time.

This works very well for most applications. For example, an LED will not light at all if you supply a low voltage. The light output is not linear to the voltage input. Turning it on and off rapidly is the best way to make it appear to be dimmed. However, in some cases, you might need to actually supply a true variable voltage from an Arduino. For example, you might be supplying voltage to a component that will burn out if exposed to high voltage, even briefly. You might also be supplying power to a device with a high inductance that might react badly to a rapidly pulsating current.

The way to provide a fairly steady voltage is to use a capacitor to charge up, and therefore hold the voltage. The circuit for this is shown in Figure 8.1.

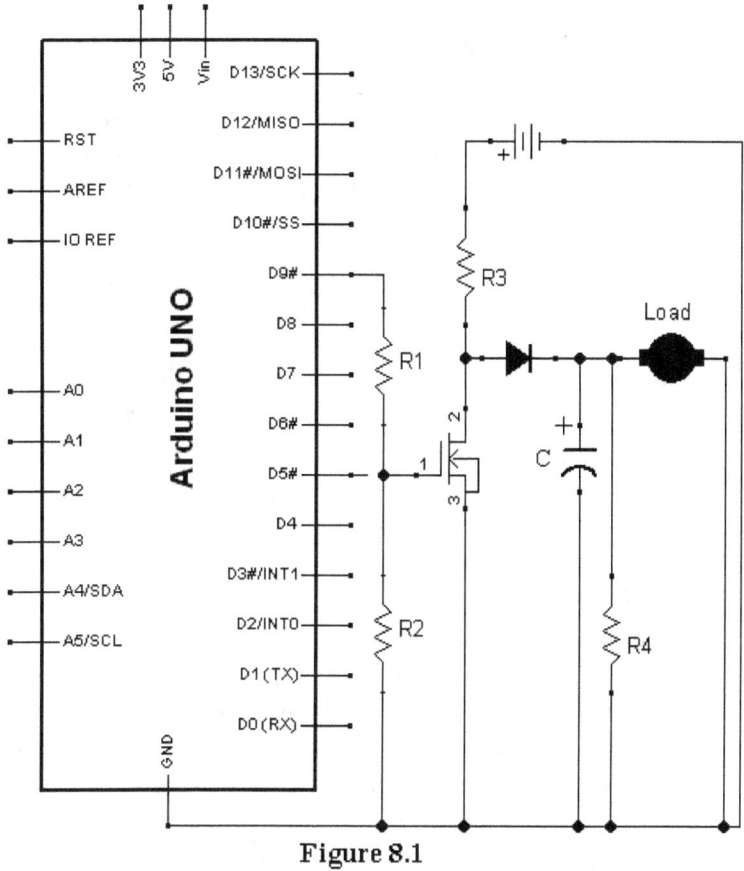

Figure 8.1

This is similar to Figure 7.1. The Arduino output controls a MOSFET. R1 is just a current limiting resistor in case something goes wrong with the MOSFET, and should be a very low value like 220 ohms. R2 should be high, like 10 K to 100 K. The values of R3 and R4 are more important. The purpose of R3 is to keep the capacitor from charging up too quickly and to prevent too much current from flowing through the transistor when it is conducting. The diode is to keep the capacitor from discharging through the transistor when it is conducting. The purpose of R4 is to allow the capacitor to discharge slightly during the time current to the capacitor is cut off when the transistor is conducting. If some of the charge on the capacitor were not discharged while current is cut off, it could still keep charging to a higher and higher voltage until it reached the full voltage of the battery. However, the load actually provides the same function. It draws current from the capacitor too. Therefore, R4 is only really necessary if the load is a very high impedance. If you can balance R3 with the load properly, you will not need R4.

Note that the analog output of this circuit is the inverse of the average output of the Arduino PWM pin. That is, the longer the PWM pin is high, the LOWER the analog voltage. Therefore, in your code, you must invert the output by subtracting it from 255. Assuming that you have a variable called value that ranges from 0 to 255 and you want the analog output to go up ad value goes up, use analogWrite(pin,255-value) to control it.

The values of R3, R4, and C depend to some extent on how rapidly you want the analog value to be able to change, and how important it is to you to have the analog output be steady with little rippling of the voltage. Lower values of C and R3 and R4 will result in faster response time of the analog output to changes in the output of the Arduino output pins. In my tests, I found 3.3 microfarads for C, 1 K for R3, and 2.2 K for R4 to work well.

Of course, it all depends on your load. You must make R3 low enough to provide current to the load, and power will be lost in R3. One solution to this is to have the analog output go to some high input device to act as an interface between this circuit and the load. Typical devices could be an op-amp, an optocoupler, or any other device that accepts an analog input and outputs a (preferably undistorted) analog output.

Chapter 9

Building the Arduino shield, components and sources.

If you are building an Arduino shield, you should generally have a good mounting that fits unto your Arduino. A good mounting should include a place to mount your components and a way of making solid connections both between the Arduino and the basic electronic components (resistors, transistors, etc.) you use in your circuits and also to the outside world, assuming that you do want to connect your device to sensors, motors, relays, and other devices. In order to do this, I recommend using a shield that connects to your Arduino with the usual pins, has a mounting space for components, and has screw connectors for outside connections as well as pin connections on the top. Figure 9.1 shows such a shield.

Figure 9.1

This particular shield can be purchased from imall.iteadstudio.com (at the time of the writing of this book). The exact link to order it at the time of this writing is http://imall.iteadstudio.com/im120417013.html and the price was $3.50, although that was a reduced price. This type of shield is generally called a protoshield or prototype shield if you want to search for one on EBay or Amazon.com.

The connections on the side are screw terminals. These are electrically connected to the pins on the bottom of the shield that insert into the Arduino and to the female wire sockets on top. This allows you to simultaneously connect your components to the Arduino and to external connections. On top of these external connectors are screws. When you unscrew the screws, holes on the sides (shown in Figure 9.2) open to allow you to insert a wire. You then screw the screws back in to lock onto the wires. This makes for a very secure connection that is still easy to disconnect if you want to.

Figure 9.2

The holes in the board are just for mounting components. You simply insert the component leads through the holes to secure them in place and tie or solder the leads from components together. You can then run wires to the female sockets to connect to the Arduino and the external connections. Figure 9.3 shows a resistor and a capacitor mounted on top and connected by a wire to the external connections.

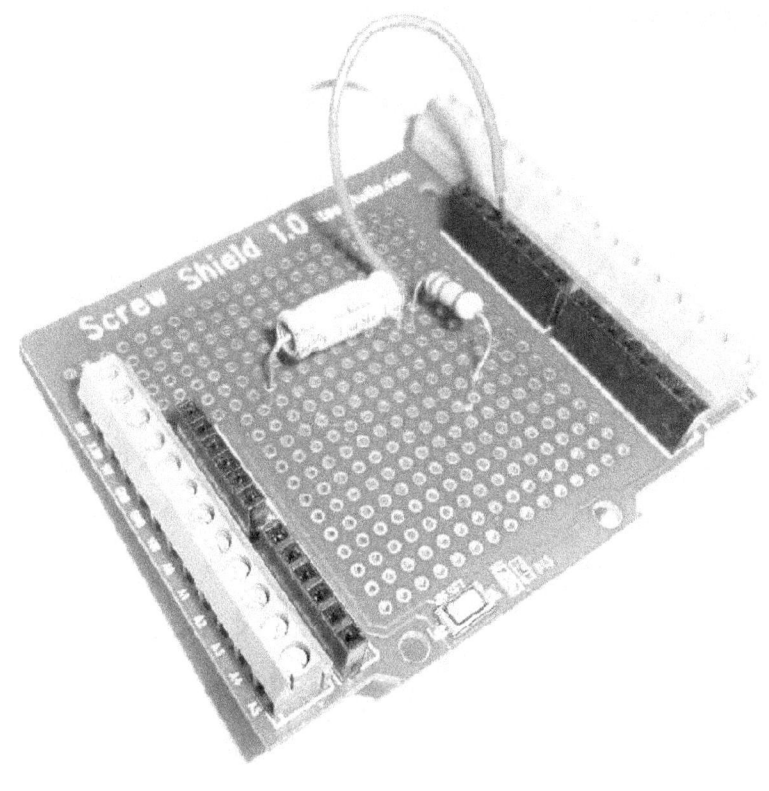

Figure 9.3

I want to make it clear that I am not saying that you absolutely have to use this particular shield. There are many other types. I simply find the screw hole attachments very convenient for securely attaching external components (e.g., sensors) while the pin holes are convenient for connecting the circuit components to the Arduino.

www.ingramcontent.com/pod-product-compliance
Lightning Source LLC
Chambersburg PA
CBHW072309200526
45168CB00014B/1163